Amazing Trains
of the World

Roland Berry and Frank Moses

Hamish Hamilton - London

First published in Great Britain 1984 by
Hamish Hamilton Children's Books
Garden House, 57-59 Long Acre London WC2E 9JZ
Copyright (text) © 1984 by Roland Berry
Copyright (illustrations) © 1984 by Frank Moses
All rights reserved

British Library Cataloguing in Publication Data
Berry, Roland
Amazing trains of the world.
1. Railroads - Juvenile literature
I. title   II. Moses, Frank
385   HE1031
ISBN 0-241-10967-1

Typeset by Katerprint Co. Ltd, Oxford
Printed in Belgium by
Henri Proost & CIE, Turnhout

Contents

Stevenson's Rocket 8
The Blue Train 9
The Mallard 10
A decorated steam train from India 12
The Deltic 13
The Union Pacific Railroad 14
Atchison, Topeka and Santa Fé Railway 15
Queen Victoria's carriage 16
A car on rails 17
The Trans-Siberian Railway 18
The Trans-Australian Railway 19
The Beyer-Garratts 20
The Black Mesa and Lake Powell Railroad 21
The TGV 22
The Romney, Hythe and Dymchurch Railway 23
The Pilatus rack railway 24
The Schwebebahn 25
The Tokaido Shinkansen Railway 26
The Evening Star 27
The London Underground 28
Freightliner 29
The Maglev 30
Robot Trains 31

8 Stevenson's Rocket

Stevenson's Rocket is probably the world's most famous steam engine. Built by George Stevenson in 1830, it became the first steam locomotive to pull an inter-city train. The original Rocket is now in the Science Museum in London; this picture is of a working replica outside the National Railway Museum in York, England. The men beside it are dressed in 1830s costume.

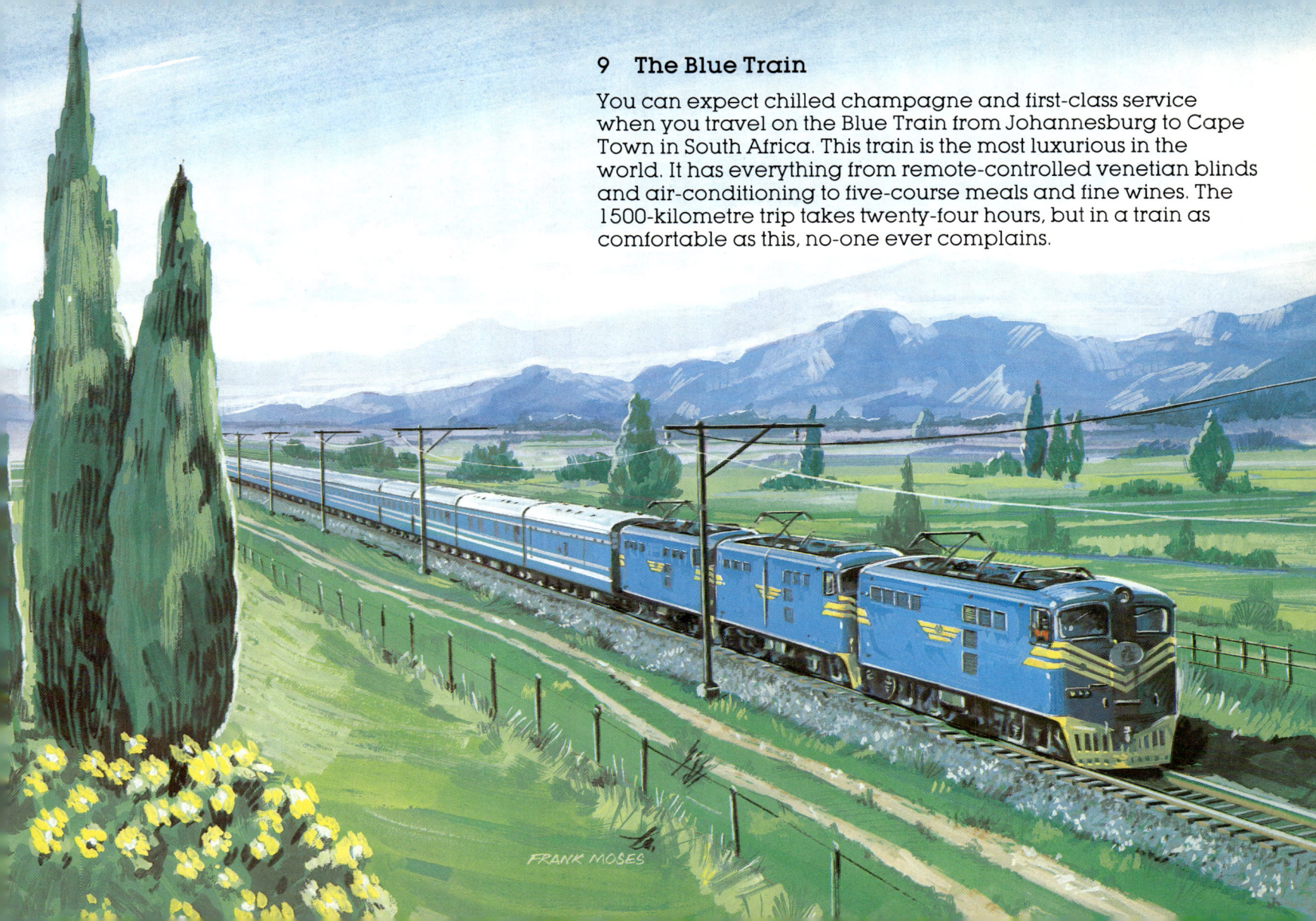

9 The Blue Train

You can expect chilled champagne and first-class service when you travel on the Blue Train from Johannesburg to Cape Town in South Africa. This train is the most luxurious in the world. It has everything from remote-controlled venetian blinds and air-conditioning to five-course meals and fine wines. The 1500-kilometre trip takes twenty-four hours, but in a train as comfortable as this, no-one ever complains.

10 The Mallard

The Mallard must be one of the most glamorous railway engines ever made. Introduced in 1935 as one of the now famous A4-class Pacific locomotives, it went on to set a world speed record for a steam train of 126 mph. Unfortunately, it blew up her engine in the process! It is now stationed at the National Railway Museum in York, England.

The picture opposite shows the driver's view of the cab.

12 A decorated steam train from India

India is one of only a few countries which still run steam trains for everyday use. Indian trains are often highly decorated, and the one shown here has won several competitions for locomotive decoration. Trains in India are often so crowded that passengers cling to the outside of the carriages and even sit on the roof.

13 The Deltic

When British Rail's Deltic diesel-electric locomotive was introduced in 1955, it was the most powerful diesel locomotive in the world. It weighed only 106 tonnes, whereas a steam engine of the same power at that time weighed over 160 tonnes.

The wheels of the Deltic are driven by electricity, generated by two huge 18-cylinder diesel engines. By 1975 each of the 22 Deltics had completed over 3 million kilometres.

14 The Union Pacific Railroad

Unlike their European counterparts, American railway-builders did not usually build railways to link existing communities, but to open up new territories and to establish new towns. Forty years after the first locomotive was delivered to the United States from Britain in 1829, the West was linked to the East by the Union Pacific Railroad. At a special ceremony at Promontory, Utah, a golden spike was driven into the track to mark the joining of the line eastward from Sacramento, California, to the line westward from the Missouri River.

15 Atchison, Topeka and Santa Fé Railway

In 1880 the first steam train puffed into the town of Santa Fé in New Mexico, America. It looked very much like the Union Pacific locomotive opposite. Nowadays, the Atchison, Topeka and Santa Fé Railway is used almost entirely for freight.

On its way to Los Angeles, the trains face some of the most difficult conditions in America. These include five mountain passes, four of which are over 2000 metres high.

16 Queen Victoria's carriage

This elaborate day saloon, upholstered in finest silks, is part of Queen Victoria's favourite railway carriage. It was built in 1869 at the joint expense of the Queen and the London and North Western Railway.

The carriage was lavishly furnished with specially built settees and armchairs, thick carpets, electric lights and also oil lamps (which the Queen preferred). The night saloon was just as luxurious and included a bed, a lavatory and washing facilities.

The carriage is still preserved, and can be seen at the National Railway Museum in York, England. But the public are not allowed inside it because the fabrics have become very fragile and could easily be damaged.

17 A car on rails

Although Peru has very varied rolling-stock, the oddest vehicle must be this modified car. Originally built to run on roads, cars like this are now used to transport railway maintenance workers along the highest standard gauge railway in the world. In the Andes Mountains, the Peruvian State Railway rises to a height of 4781 metres.

18 The Trans-Siberian Railway

From Moscow to Vladivostock the Trans-Siberian Railway runs through 9,438 kilometres of some of the world's most varied countryside, from dark forests and snowy tundra to hot desert and empty plains. It is the longest run in the world, taking eight days, four hours and twenty-five minutes. During this time, it makes ninety-seven stops. Although the engines are now diesel, the picture shows one of the older steam trains struggling through the snow, so often a feature of a journey on this railway.

19 The Trans-Australian Railway

A trip on the Trans-Australian Railway from Sydney to Perth takes three days and covers 3,959 kilometres. During 478 of these kilometres the track runs dead straight across the Nullabor Plain, a completely flat area where nothing grows but small bushes. In the Aborigine language, 'Nullabor' means 'without any tree'. The train in the picture is hauled by three Indian Pacific diesel-electric locomotives.

20 The Beyer-Garratts

This locomotive was owned by the East African Railway and was one of many similar steam engines built by Beyer-Garratt. These engines were used extensively in Africa, and were among the most powerful ever built.

The design of the engine had many peculiar features, the most obvious being that it was in fact two engines — one at either end of the locomotive with a boiler in the middle. Because the engine was articulated, the locomotives could get around the tight bends on the narrow metre-gauge track.

21 The Black Mesa and Lake Powell Railroad

The 2.31 million kilowatt electricity generating station in Navajo, Arizona, USA, is supplied with fuel from the Black Mesa coal mines 124 kilometres away. A power station of this size needs over 20,000 tonnes of coal a day, which is delivered by huge trains, each pulled by three of these Co-Co General Electric E60C locomotives. The power used to pull the trains is roughly equal to the amount of electricity needed to light 126,000 lightbulbs.

22 The TGV (Train à Grande Vitesse)

France's TGV currently holds the record as the world's fastest passenger train, having reached a speed of 270 kilometres per hour between Paris and Lyon. On a special test run, it exceeded speeds of 380 kph. Although not a luxurious train like the Blue Train (p. 9), the TGV has some of the most advanced safety devices now available. One of these is a control which senses gradient changes and adjusts the amount of electric power needed to keep the train running at the same speed.

23 The Romney, Hythe and Dymchurch Railway

This miniature railway claims to be the smallest public railway in the world. Opened in 1927, it covers the 22 kilometres between Hythe and New Romney in Kent, south-east England. Each train is one-third the size of a normal train.

This engine is called Doctor Syn. It is one of two such engines owned by the railway. Both were modelled on a Canadian Pacific engine of the Pacific Class, and were built by copying a photograph of the original engine.

24 The Pilatus rack railway

Climbing up the side of Mount Pilatus in Switzerland, overlooking Lake Lucerne, is the steepest railway in the world. Unlike most railways, a railway like this has a rack of steel teeth between the rails which the large cog wheels lock onto – a bit like climbing a ladder. This complicated arrangement also stops the train from slipping back down the steep track.

25 The Schwebebahn

The German town of Wuppertal is the home of the world's first commercial mono-railway, the Schwebebahn. Operating much like a bus service, the Schwebebahn travels above the city streets and over the River Wupper, stopping frequently at the many stations. Since the railway was opened in October 1900, the trains have been redesigned four times. This blue and orange type is the latest.

26 The Tokaido Shinkansen Railway

In their first 20 years of service, the Tokaido Shinkansen trains have carried over 2000 million passengers and travelled the equivalent of 900 round trips to the moon and back.

Nicknamed the Bullet, a train like this can reach speeds of up to 210 kph. It travels across the whole of Japan, linking the islands with long bridges up to nearly four kilometres long. Earthquake sensors are installed at points along the route to warn drivers of any earthquake tremors.

27 The Evening Star

The Evening Star was the last steam locomotive built for British Railways, and was completed in 1960 in Swindon, Wiltshire. Sadly these magnificent locomotives can now only be seen in museums and on private lines run by enthusiasts.

There are many reasons for the disappearance of steam as a source of power. Two of the most important ones are that steam engines do not convert fuel into power efficiently, and that they require more people to run than do electric and diesel trains.

28 The London Underground

The London Underground, known locally as the Tube, is the most extensive underground railway system in the world. It serves 277 stations and covers 418 kilometres, of which 162 kilometres are underground. Some sections of the track go as deep as 67 metres. Since the first section opened in 1863 many changes have been made to the trains and other equipment. This picture shows the latest information boards. These not only tell you which route the next train is taking, but also how long you will have to wait for it to arrive.

29 Freightliner

When Freightliner was introduced in 1965, it meant that all kinds of goods could be transferred quickly and easily from one type of transport to another. The goods are loaded into large containers, designed to fit both onto flat-bed rail cars and onto lorries. When the train or lorry has reached its destination, a huge gantry crane lifts off the containers.

30 The Maglev

One of the most exciting advances in recent times has been the development of 'magnetically levitated vehicles', known as Maglevs.

These vehicles hover above a single rail, being held up and propelled by electro-magnetism. You can see the row of electro-magnets placed along both sides of the monorail in this picture. Incredible speeds have been achieved with this system. The record is held by this vehicle in Miyazaki, Japan, which reached 376 kph.

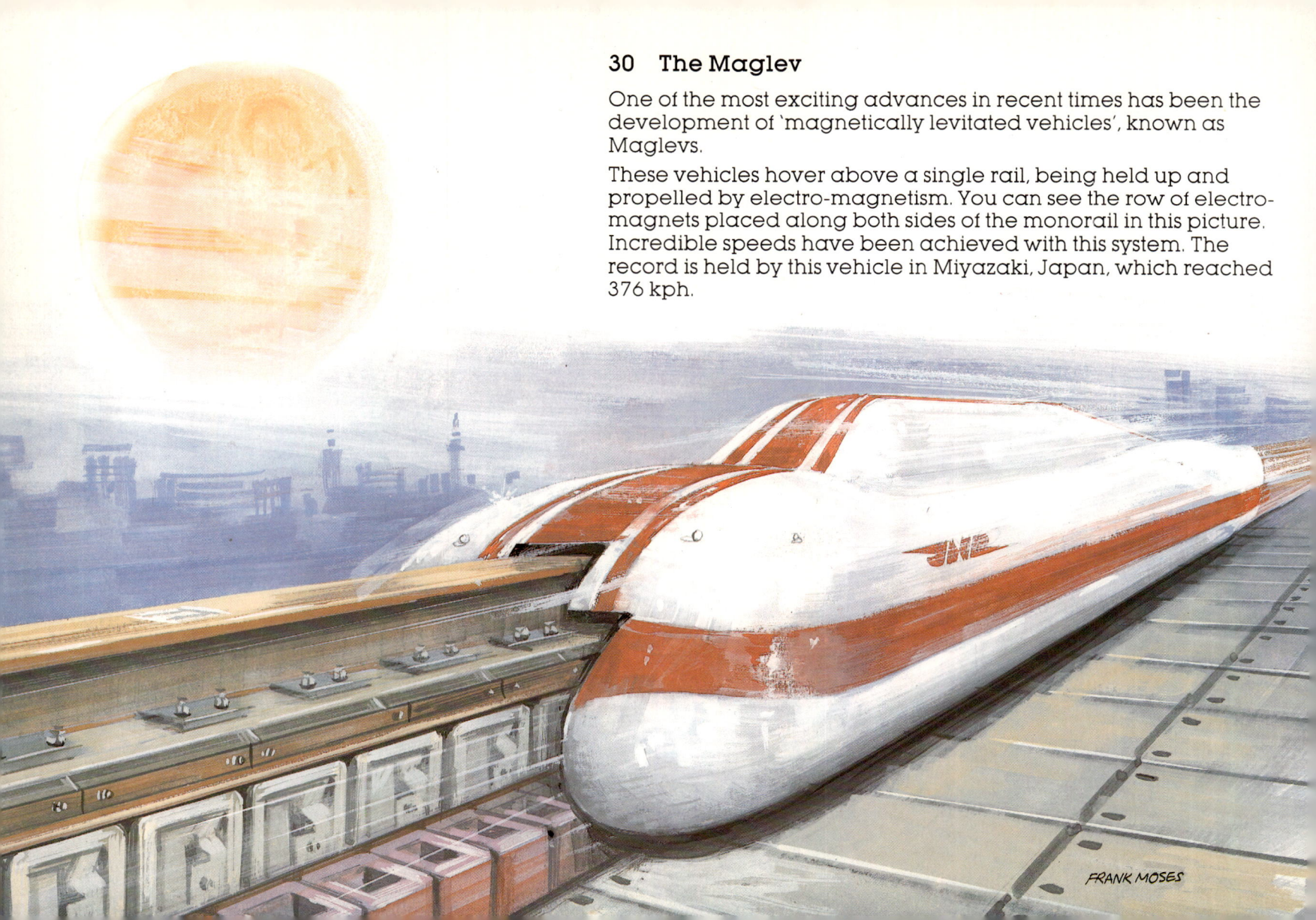

31 Robot trains

These electric trains are Europe's first 'robot' trains. They are used to shuttle passengers to and from Gatwick Airport near London. They are operated from a central control room, so there is no need for a driver in the train itself. Although power for the train comes from a single rail, the carriages run on a concrete track, their rubber-tyred wheels making the journey quiet and comfortable.

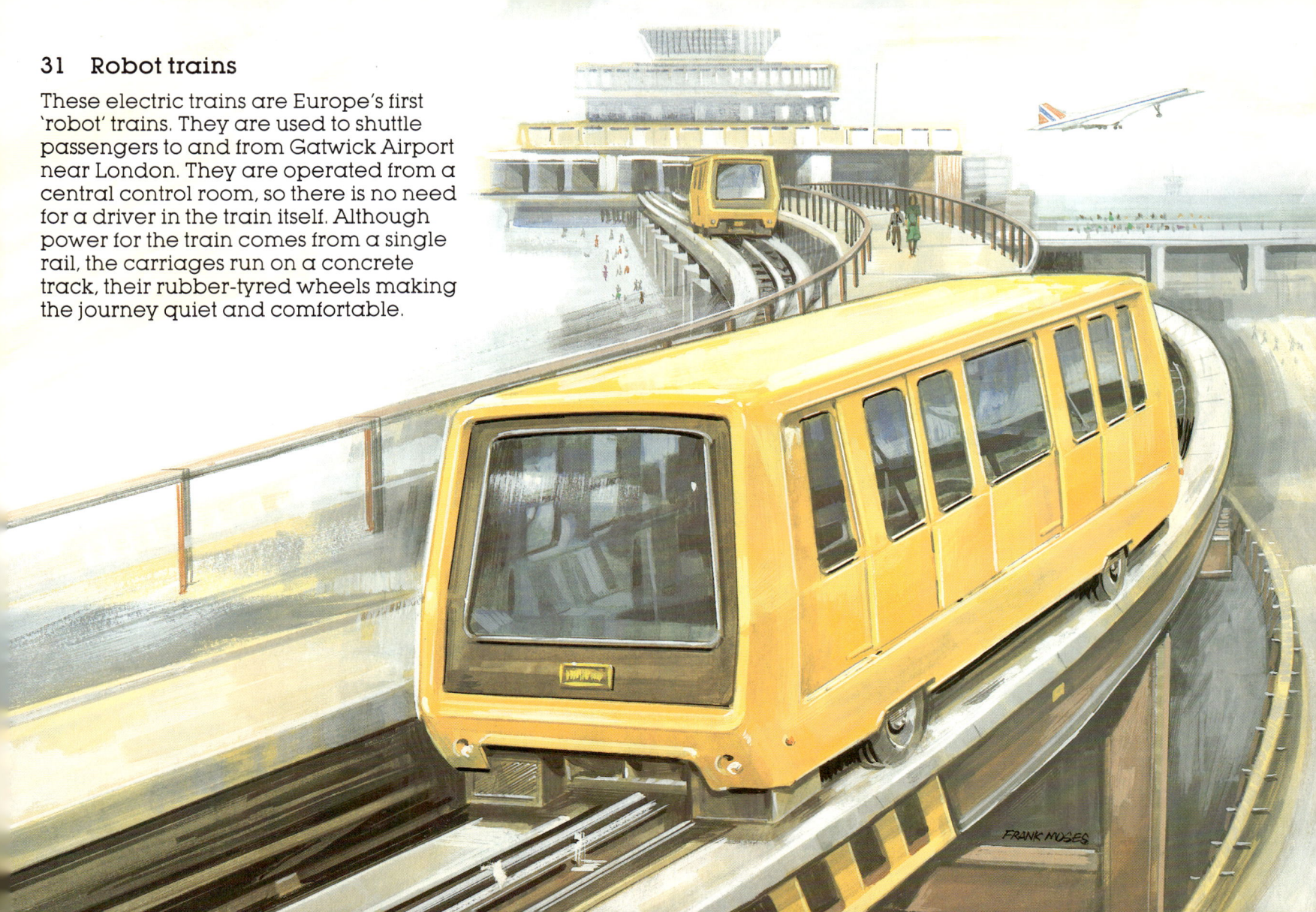